S0-BOB-232

SECRETS OF THE RAINFOREST

PREDATORS AND PREY

BY MICHAEL CHINERY

🌳 CRABTREE

Crabtree Publishing Company

PMB 16A, 350 Fifth Avenue
Suite 3308
New York, NY
10118

612 Welland Avenue
St. Catharines, Ontario
Canada L2M 5V6

Created by
Cherrytree Press

© Evans Brothers Limited 2000

All rights reserved. No part of this publication may be reproduced, stored in a retrieval system or transmitted in any form or by any other means, electronic, mechanical, photocopying or otherwise, without prior permission of the publishers and the copyright holder.

Library of Congress Cataloging-in-Publication Data

Chinery, Michael.
    Predators and prey / by Michael Chinery.
        p. cm.– (Secrets of the rainforest)
    Summary: Describes some of the many different animals, from
anteaters to the great cats, that prey on other creatures in rainforests.
        ISBN 0-7787-0227-8 (pbk) – ISBN 0-7787-0217-0
1. Predatory animals–juvenile literature. 2.Predation (Biology)–Juvenile
        literature.[1. Predatory animals.] I. Title.
    QL758 .C45 2000
    591.5'3–dc21

                                                LC 00-020390
                                                        CIP

Co-ordinating Editor: Ellen Rodger

Designed and produced by A S Publishing

Editor: Angela Sheehan
Design: Richard Rowan
Artwork: Malcolm Porter
Consultant: Sue Fogden

Acknowledgements
Photographs: *All by courtesy of Michael & Patricia Fogden*
with the following exceptions: BBC Natural History Unit
14 top, 16/17 bottom, 22, 22/23 top, 23 bottom, 24
bottom, 26/27 bottom, 27 right, 28/29, 29

1234567890 Printed in Hong Kong by
Wing King Tong Co. Ltd 543210

◦ CONTENTS ◦

☼ PREDATORS AND PREY ☼

Rainforests grow in warm, wet regions near the equator. They are dense evergreen forests that receive at least 80 inches (2 meters) of rain each year. Rainforests have a greater variety of plants and animals than anywhere else on earth. Some animals that live in rainforests are **herbivores**, or planteaters. They depend on plants for food. Some animals are **carnivores**. Carnivores are **predators** who hunt other animals for meat. Other animals are **omnivores** who eat both plants and animals.

FOREST FOOD CHAINS

The eagle that swoops down and snatches a monkey from the tree-tops is a carnivore.

AMBUSH

SITTING motionless with its front legs folded up under its head, this mantis (below) looks more like a withered flower than a hungry killer. It goes unnoticed by other insects, but it is alert and waiting to pounce. When another insect comes within range, the mantis shoots out its spiky front legs and its victim is hopelessly trapped.

Gladiator spiders also **ambush** their prey, using an unusual kind of web that acts as a net.

Gladiator spiders hide, often close to the ground, and wait for insects to pass below them. The gladiator spider (right) holds a small but stretchy net between its four front legs. When prey comes within range, the spider spreads its legs to expand the net and then drops it over the victim.

The meat that the eagle eats is created from the plants that the monkey has eaten. This cycle of eating and being eaten is an example of a food chain. All animals belong to a food chain. Some **food chains** have four or five links, called webs, but all chains start with plants.

CHASING AND TRAPPING

Animals that catch and eat other creatures are called **predators**. The animals they catch are their prey. Predators have two main ways of catching food: they chase it, or they lie in wait for it and take it by surprise. Spiders and a few other animals make traps to catch their prey.

Most predators in the rainforest are hunters. They see, smell or hear their prey from some distance away, and then go after it. Some, such as the chameleon, approach their prey silently and slowly. Others, such as the eagle, strike swiftly. Often, predators use a combination of the two. Cats and snakes approach silently before making a final strike.

◀ Silence, concentration, speed, and strength are the predator's tools. At the right moment, this jaguar will run its victim down.

● INSECTS AND INSECT-EATERS ●

THE MOST numerous creatures in rainforests are **invertebrates**, including insects and spiders and other small animals. Many of them, such as butterflies and moths, are planteaters. Others, including cockroaches, live as scavengers, eating dead leaves and other debris on the forest floor. Some are meateaters. All are food for larger animals, including birds, **amphibians**, **reptiles**, and **mammals**.

Ants are everywhere on the ground and in the rainforest trees. More than 400 different kinds of ants live in the Amazon forest of South America. They feed on plants and animals and they, in turn, are eaten by many other animals.

ANTS ON THE MARCH
Army ants live in the **tropics** of Africa and America. They live in enormous colonies, some of which contain as many as 20 million ants. Army ants feed mainly on other insects, but no animal is safe from them.

▲ Butterfly eggs and caterpillars are food for huge numbers of birds and other predators. Very few caterpillars survive to grow up into new butterflies.

▼ These army ants are transporting a wasp larva back to their nest. They bring food home for the queen and the thousands of workers that look after her as she lays her eggs.

▲ Termites have many enemies, but predators are often attacked by the large-jawed soldier termites when they break into termite nests.

▲ A white-plumed antbird perches close to columns of army ants and eats cockroaches and other insects as they scurry away from the ants.

◀ Tree frogs eat large numbers of insects in rainforests.

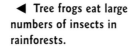

Army ants will eat anything that cannot escape, including large snakes that have eaten big meals and cannot move quickly. Thousands of ants with sharp jaws quickly reduce their prey to skeletons.

MAKING CAMP

Army ants do not build permanent homes. If these huge colonies stayed in one place, they would soon run out of food. Instead, they 'camp' for a while, and then move on to new hunting grounds.

American army ants make their camps, or bivouacs, in sheltered spots, often between the **buttress roots** of large trees. Thousands of worker ants link their legs together to make a ball about three feet (one meter) across. The queen and thousands of other workers that look after her, rest in this living tent at night. At daybreak the workers form into columns and march off to find food. Large ants called soldiers, march at the edges of the columns. The ant column catches and kills anything in its path. They eat some of the food as they catch it, but store the rest by the sides of their paths and collect it on the return journey.

TERMITES

Termites are small, soft-bodied insects that live in large colonies. Many of them nest in trees or on the ground. Some termite species build enormous mounds that house several million insects. Termites are often mistaken for ants, although the two groups are not related. Ant-eating mammals usually eat termites as well. Termites do not have stingers and their soft bodies are easier for predators to digest.

EATING ANTS

Although each ant is small, nests or colonies may contain many thousands of them. Predators eat a lot of ants by breaking into a nest. Some rainforest mammals are specialists in eating ants and termites. They usually have big claws and long snouts to rip open and poke into nests. Long sticky tongues help them lick up the ants.

GIANT ANTEATER

The giant anteater lives on the plains and in the forests of tropical Central and South America. It feeds on large ants. It finds ant nests with its excellent sense of smell. Its tongue is about 24 inches (60 cm) long and covered with thick saliva that traps the ants and prevents them from stinging its mouth.

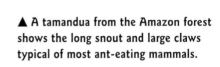

▲ A tamandua from the Amazon forest shows the long snout and large claws typical of most ant-eating mammals.

◀ Tamanduas are small tree-dwelling anteaters. They move slowly through the branches at night, clinging on with their tails as well as with their claws. They feed on termites as well as ants. This one is a northern tamandua from Central America.

▶ Clinging to a woody liana plant with its back legs and long furry tail, this silky anteater is ready to defend itself against eagles and other predators with its big front claws.

Anteaters have no teeth and they grind up their food with powerful stomach muscles. An anteater needs to eat several thousand large ants every day, but it does not take them all from one nest. It visits many nests in a day and takes no more than a few hundred ants at a time. In this way, the anteater does not do too much damage to the colonies and assures that there will be food for another day.

SCALY ANTEATERS

Pangolins are called scaly anteaters because they are covered with tough, overlapping scales. Several species live in the rainforests of Africa and in Asia.

All pangolins are excellent climbers. They use their large claws and the sharp edges of their scales to grip the tree trunks as they climb. Some pangolins can hang by their tails. The giant pangolin, which is about 60 inches (150 cm) long and weighs about 66 pounds (30 kg), lives on the ground and can eat many thousands of insects in one night. Thick mucus in its mouth and throat prevent it from being stung or bitten. A pangolin can roll itself into a scaly ball, and it is then safe from almost any large predator.

❖ FEATHERED PREDATORS ❖

THOUSANDS OF different kinds of insects and small creatures live in the rainforests and are food for hundreds of different birds. The birds' beaks are adapted for catching insects either on the ground, in the trees, or in the air.

Although there are not many small plants on the forest floor, there are always plenty of dead leaves and branches that have fallen from the canopy. There is also plenty of fallen fruit. These all provide food and shelter for worms, beetles, woodlice, slugs, snails, millipedes, and other small scavenging animals. The scavengers are eaten by spiders, centipedes, carnivorous beetles, and other small predators. All of these small creatures make the forest floor a great hunting ground for birds such as pittas and antbirds.

ANTBIRDS AND PITTAS

About 250 species of antbirds live in South and Central America. Not all of them live in the rainforests.

◀ This blue-crowned motmot has caught a tasty frog to feed its babies. It often kills its prey by crushing it on a branch before eating it.

▶ The small eyes of this scale-backed antbird are always on the lookout for ants and other juicy insects.

KINGFISHERS THAT DON'T FISH

KINGFISHERS do not all eat fish. Many members of this large family live and feed in the rainforests. The shovel-billed kingfisher lives in New Guinea. It has a massive beak like a shovel which it uses to dig for worms and other invertebrates in muddy ground. It also catches crabs in **mangrove** swamps. Paradise kingfishers also live in New Guinea and neighboring areas. They perch on low branches and fly quickly to the ground or to nearby leaves to catch lizards, worms, or insects. The buff-breasted paradise kingfisher (left) lives in northern Australia.

◀ Trogons spend a lot of their time sitting on leafy branches. They fly out from their perch with their beaks wide open to scoop up insects in mid-air. This orange-bellied trogon lives in the rainforests of Central America.

They are round-bellied birds with long legs. Almost all antbirds live on or close to the ground. Their short, rounded wings are ideal for flying through dense vegetation, although they are not strong fliers. Large flocks, often containing several different species, move over the forest floor as they search for ants and other small animals.

Pittas are small, colorful birds, with short stubby tails. About 30 species live and feed on the rainforest floor. Apart from the African pitta, they all live in Southeast Asia and on islands in the South Pacific. Most of them fly very little. Instead, they run around scratching in the leaf litter for worms and insects, especially termites.

▶ Standing on top of rotting leaf litter, this giant pitta has just finished feeding on rainforest ants.

WELL-NAMED WOODPECKER

Woodpeckers find their food by pecking into tree trunks with their beaks and dragging out insects that feed there with their long tongues. Woodpecker beaks are long and sharp and they can hammer them into the trees as fast as twenty times a second. Specially formed feet, with two toes pointing forward and two pointing back, allow the birds to cling firmly to the trunks while they are hammering.

SUNBITTERNS AND POTOOS

The sunbittern lives in tropical Central and South America, usually near water. It snaps up insects and spiders among the decaying leaves on the rainforest floor and catches frogs and crustaceans by jabbing its beak into pools of water. Sunbitterns are about 20 inches (50 cm) long, but are difficult to see when resting on the ground. When it is threatened by a predator, the sunbittern spreads its wings to look bigger and scarier. Its wings have large eye-like markings on a background of golden feathers.

The potoo is a bird that is even more difficult to find in the dense rainforest. At night, it flies and catches moths and other large insects. During the day the potoo sits motionless and looks like a broken branch.

▼ The sunbittern can spread its wings several sizes larger than its body when it is threatened.

▲ A black-cheeked woodpecker returns to its nest hole with some grubs that it has pecked from a tree trunk.

▲ Is it a bird or is it a branch? The potoo (above) sleeps like a log and keeps so still that other animals pass it by.

▲ The boat-billed heron (above right) lives close to rivers in the rainforests of Central and South America. It comes out to hunt as the sun goes down and scoops up fish in its huge bill.

A NOSE FOR FOOD

Most vultures live in open country, using their keen eyesight to spot dead and dying animals on the ground far below them. The king vulture of the Central and South American tropics is one of the few **carrion-eating** birds living in the forests. Unlike most other birds, it has a good sense of smell. It is not easy for birds to see carrion on the forest floor while they are flying in or above the canopy, so they need to be able to smell it. As well as eating dead birds and mammals, the king vulture feeds on fish stranded on the banks of rivers and lakes after a flood.

▲ The pauraque lives in Central America and hunts at night. It is almost invisible when sitting in the leaf litter by day.

▶ The king vulture's beak is made for ripping decayed flesh. It eats the meat of animals that are already dead. The skin on its bill is called a wattle.

❖ COLD-BLOODED HUNTERS ❖

SNAKES, LIZARDS and crocodiles are reptiles. Many different kinds of reptiles live in the rainforests. Many of them climb trees. Reptiles are often described as **cold-blooded** animals, but they are not always cold. Their bodies stay about the same temperature as their surroundings. In cold weather they cool down and become very slow, but in warm weather their bodies warm up and they move quickly.

Snakes, lizards and crocodiles use the color of their bodies to help them find food. Most rainforest snakes and lizards have skin that **camouflages** them, or blends in with the vegetation or the forest floor. Some lizards, such as chameleons, can change their skin color to match their backgrounds. Camoflauge allows them to wait and watch without being seen by their prey.

BOA FAMILY
Rainforests are home to the world's biggest snakes. The anaconda from South America grows to about four feet (ten meters). The anaconda belongs to a group of snakes known as boas.

ANACONDA

The anaconda is rarely found far from rivers and spends much of the daytime basking in the shallow waters. It feeds mainly at night but it is too heavy to chase things, so it usually lies in wait on the river bank. Its prey includes fish, birds, mammals, and even crocodiles. Mammals are also caught when they come to the river banks to drink.

The anaconda is not a poisonous snake. It either suffocates its victims by wrapping its huge body around them or it drags them into the water and drowns them. It easily overcomes tapirs and capybaras – pig-sized relatives of the guinea pig – and then swallows them whole. The snake in the photo (left) is more than 13 feet (four meters) long.

▲ This scrub python from the rainforest of northern Australia is not easy to spot hidden among the dead leaves on the forest floor. It can slither quietly up to its prey without being seen.

BOA'S TIGHT SQUEEZE

Several other boas, including the boa constrictor, live in South America. The boa constrictor feeds on birds and mammals. It kills its prey by **suffocating**, dragging, or drowning it. The emerald tree boa lives mainly in the trees, where it drapes itself over a branch and waits, camouflaged, until birds or monkeys come within its range. When its prey is near, it strikes rapidly, grabbing the animal in its mouth while hanging on to the tree with the rest of its body. Although most tree-dwelling snakes detect their prey by sight, the tree boa's mouth is surrounded by heat-detectors that tell the snake when a **warm-blooded** animal approaches.

◄ A boa constrictor drapes its powerful coils over a buttress root. Boas live for months without food and sleep for days after a big meal. They can kill and swallow animals much larger than their own heads.

► The emerald tree boa lives in the rainforests of Central and South America. It keeps its head uncovered and uses heat-detectors around its mouth to tell when its prey is approaching.

LIZARDS IN THE TREES

Hundreds of different kinds of lizards live in rainforest trees. They climb the trees using their sharp claws. Some lizards can even glide from tree to tree. The big iguanas that live in the rainforests of Central and South America feed mainly on flowers and fruits, but most lizards are predators. Their prey ranges from insects to deer and other mammals. They detect prey by sight and smell, continually flicking their tongues to pick up scents from their environment.

CAREFUL CHAMELEONS

Chameleons are colorful lizards that move slowly through the rainforest trees. They rarely move more than one leg at a time. Each of their feet has toes that grip the branches like a pair of tongs. Chameleons grip the branches with their tail, something that no other lizard can do.

Chameleons are able to change color to adapt to their environment. They also have excellent eyesight. Their eyes move in all directions and work independent from each other. This allows the lizard to look backwards with one eye and forwards with the other. When a chameleon spots an insect, it moves slowly toward it. Both eyes turn to focus on the prey and then the chameleon quickly shoots out its tongue. The chameleon's tongue is elastic and it can stretch to twice the length of its body. Prey is caught on the sticky tip of the tongue and pulled into the mouth.

▼ The green anole lizard blends with the forest leaves and is hidden from predators as well as prey. It lives in the rainforests of Central and South America.

▲ This spiky lizard is called a forest dragon. It looks fierce, but it is harmless to people. It feeds on insects and other small animals.

▼ You have to be quick to catch a fly. The chameleon completes the operation in under a tenth of a second.

CLINGING GECKOES AND ANOLES

Geckoes are famous for their ability to walk upside down. The underside of their broad toes have tiny hairs with tips that act like suction cups. Geckoes can cling to almost any surface – including shiny wet leaves. Most geckoes are nocturnal and they feed mainly on insects, which they find by sight rather than by scent. They do not flick out their tongues like other lizards.

Unlike most lizards, geckoes have voices. The tokay gecko of Southeast Asia is up to 14 inches (36 cm) long and is a very noisy animal. When it is threatened, it opens its big mouth and barks like a dog.

Anoles are small lizards. Most anoles live in trees. Like geckoes, they have suction pads on their toes that allow them to cling to smooth surfaces. Anoles are omnivorous. They eat both fruit and insects. Some can change color according to the temperature and the amount of sunlight that hits their bodies.

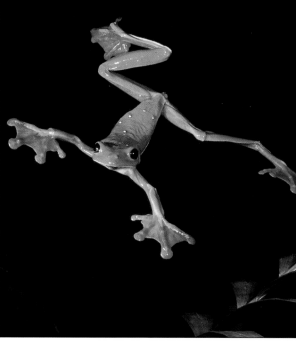

▲ Flying frogs live in the rainforests of Southeast Asia. Their large webbed feet act like a little parachutes, helping them to glide up to 49 feet (15 meters).

FLYING LIZARDS

Flying dragons live in the rainforests of Southeast Asia. They catch and eat insects on tree trunks and branches. Although they cannot really fly, they can glide as far as 65 feet (20 meters) from tree to tree. They use colorful flaps of skin supported by extra-long ribs as wings. In the air they are bright enough to be mistaken for birds or butterflies, but as soon as they land on a tree trunk they fold their wings away and melt into the background. This helps them catch their food and avoid becoming food for other animals.

▼ This gecko has broad toes. Its large eyes help it to find insects at night.

► In full flight, this flying dragon shows the delicate ribs that support its colorful wing flaps.

FROGS WITH BIG APPETITES

. . . .

FROGS belong to the group of animals called **amphibians**. Frogs are cold-blooded, but unlike reptiles, they do not have scaly skins. Hundreds of different kinds of frogs live in the rainforests, either on the ground or in the trees. They feed on all sorts of insects, and some of them have such enormous mouths that they can swallow mice and other frogs. Frogs often use their front feet to push their prey into their mouths. Large meals have to be helped on their way down the frog's throat, and the frog does this by pushing with its eyeballs! The horned frog (right) lives in the Amazon rainforest. It is swallowing a lizard.

The parachute gecko has a loose fold of skin on both sides of its body, and this opens up to form a parachute when the animal stretches its legs. It also has webbed feet and a flap all round its tail, and these all help to keep it airborne. The parachute gecko is not as good a glider as the flying dragon, but if it misses its target, it simply drifts to the ground and then climbs up another tree.

RUNNING ON WATER

. . . .

Basilisk lizards live in dense waterside vegetation in the forests of tropical Central and South America. They drop into the water when alarmed by a predator and can actually run across the water surface on their long back legs. As long as they move quickly, their long fringed toes keep them up. If they slow down they sink into the water and then have to swim. They can also rear up on their back legs and run across the ground. Basilisks eat a mixture of plant and insect food and are eaten by large birds and caimans.

❖ DANGER IN THE WATER ❖

THE RIVERS THAT run through the rainforests look cool and calm but they are teeming with life. When the rivers flood, many of the animals that live in the river swim among the trees and capture land animals that are unable to escape the rising waters

Archer fish live in the **mangrove swamps** of Southeast Asia and northern Australia. They catch most of their food in the water but they also eat insects that live near the banks of rainforest rivers. Archer fish capture insect prey by hitting them off of overhanging plants with 'bullets' of water shot from their mouths.

QUICK CAIMANS AND CROCODILES

Caimans are reptiles that belong to the same family as crocodiles and alligators. They live in South America and spend a lot of time sunbathing in groups on the muddy banks of rivers. Caimans hunt by sight, catching their prey either in the water or on land. They are large, agile, and fast animals that swim close to the banks and often tackle large mammals, such as pigs that have come down to drink. Caimans also eat fish, frogs, turtles, and birds. They in turn, are the prey of jaguars and large snakes.

▲ Two water bugs fight in the water.

▼ The dense vegetation (below left) typical of the rainforest edge is mirrored in this quiet lake. Underneath the water there are many predators.

▲ Piranhas are flesh-eating fish that attack in packs. This South American black caiman (above right) has just caught a large piranha.

▼ The eye-lash viper is a venomous snake that hunts close to the water. It feeds mainly on birds, which it catches when they come to drink.

CRUISING CROCODILES

Crocodiles also live in rainforest rivers and hunt for prey along river banks. Like caimans, they eat fish, turtles, and large mammals they find drinking water from the river.

WHAT A SHOCK!

The electric eel lives in the rainforest rivers of South America. It is up to 9.8 feet (3 meters) long and its body contains special muscles that generate electricity. These muscles, which account for up to half of the fish's weight, also store electricity like batteries. The fish electrocutes its prey by firing electric currents into the water. It feeds mainly on other fish and frogs, but it can make enough electricity to kill a horse standing in the water. The eel usually lives in stagnant backwaters that do not contain much oxygen. It sometimes comes to the surface for air.

RAZOR TEETH

About 20 species of piranha fish live in the rivers of Central and South America. Piranhas feed mainly on other fish. They are ferocious fish and will attack almost any animal that enters their water. Although each fish is no more than 24 inches (60 cm) long, they swim in large groups, called shoals. Their razor-sharp teeth carve slices of flesh from their prey and a shoal can reduce a large animal to a skeleton in a few minutes. Piranhas hunt mainly by scent and get excited when they smell blood in the water. Wounded animals are almost certain to be attacked.

❂ TREE-TOP CARNIVORES ❂

MOST OF THE planteating animals of the rainforest live in the canopy. Many rainforest predators also live in the canopy, or the understorey, the layer of smaller trees just below the canopy. The canopy is where the tree tops join together. It is rich in food, such as seeds and fruit. Canopy predators are small and **agile**. Most of the carnivorous mammals living there are nocturnal. They have large eyes that help them find their food and judge distances when leaping from branch to branch.

A variety of small cats, some no bigger than **domestic** cats, live in the rainforests. Several have been hunted so much for their fur that they are threatened with **extinction**.

BIG EYES AND SHARP EARS

Most forest cats hunt at night, but even those that hunt in the daytime need big eyes to spot their prey in the dark, leafy rainforest. They also have excellent hearing and a good sense of smell. When they have spotted their prey, the cats slink quietly and slowly towards it and then pounce on it when they get close enough. They eat almost anything they can catch, but birds and small mammals, including monkeys, make up most of their food.

▶ Margays hunt mainly by day and catch monkeys and birds in the forest canopy of South America. They can climb head-first down a tree trunk almost as easily as they can climb up.

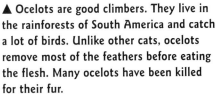

▲ Ocelots are good climbers. They live in the rainforests of South America and catch a lot of birds. Unlike other cats, ocelots remove most of the feathers before eating the flesh. Many ocelots have been killed for their fur.

◀ Clouded leopards live in the rainforests of central Asia. They climb well and can even crawl upside down along branches to reach their prey. They eat birds, monkeys, deer and pigs. Clouded leopards kill larger animals by dropping on them from a branch.

▶ Fishing cats live in mangrove swamps and other wet forests. They catch fish and also eat frogs, crabs, insects, and birds.

LINSANGS AND GENETS

Linsangs and genets and other members of the mongoose family hunt in the rainforests of Africa. The African linsang looks like a slender cat. It can run along the slimmest branches of trees in the rainforest canopy. Genets are just as quick. Both animals are **nocturnal** and rarely come down from the trees. Lizards, birds, their eggs and young, and small mammals are their main prey.

CLIMBING COATIS

The ringtail coati, sometimes called the coatimundi, is a member of the racoon family. It lives in the forests of South America. The ringtail coati is a good climber. It finds most of its food by rooting around in debris on the forest floor. It sometimes eats lizards and mice. Beetles, ants, spiders, and other **invertebrate** animals make up most of its diet, together with fruit, which it gathers up from the ground or picks fresh from the tree-tops.

▲ A white-nosed coati raids a banana tree. Coatis live in the South American rainforests and eat meat and fruit.

AGILE BUSHBABIES

Bushbabies and lorises are distant cousins of monkeys. They look like small monkeys and often live with them in the forests of Africa and Asia. Monkeys roam the forests by day but bushbabies and lorises come out only after dark. They have big eyes that help them to see at night and catch food, but they rely more on sounds and smells to locate their prey. They eat insects and other small animals, fruit and flowers.

Bushbabies are agile animals. They use their long, strong back legs to leap from tree to tree. Their long, bushy tails help them to steer and to keep their balance. Bushbabies are so quick that they can catch moths in flight by clinging tightly to a branch with their back feet and shooting their arms out to catch the insects in their paws.

SLOW LORISES

Lorises move very slowly. They have short tails and cannot leap from trees. Lorises rarely take more than one foot off a branch at a time. They find their prey by scent, and creep slowly up to it and grab it with their teeth or hands. Lorises eat many prey that other animals do not eat, such as millipedes and caterpillars with stinging hairs. Sometimes they rub off the hairs before swallowing the flesh.

LONG-TAILED TARSIERS

Tarsiers live on various islands in Indonesia and the Philippines. They look like small bushbabies, but they have slender tails and their back legs are nearly twice as long as the body. The tarsier's huge eyes and big ears help them find prey at night. They hunt in trees or on the ground and usually capture their prey by leaping on to it and pinning it down with their long fingers. Insects are their main prey, but tarsiers also eat lizards, birds, and bats. They make enormous leaps from tree to tree, pushing off with their powerful back legs and judging the distances accurately with their big eyes. They use their long tails as rudders.

▲ Tarsiers can leap long distances between trees. They use their tails as rudders. From a lookout post in a tree, they survey the area for prey.

◄ As well as good sight, bushbabies have sensitive noses that can pick up the scent of food in the still air of the forest.

❋ MIGHTY HUNTERS ❋

MOST OF THE animals that live in rainforests are the prey of other animals higher up in the food chain. Big cats and eagles are at the top of the food chains. They are less numerous than smaller animals but are so strong that they have no natural enemies except for humans.

GREAT CATS

Three of the world's big cats live in rainforests, although they live in other habitats as well. Jaguars live in South America, tigers in Asia, and leopards in Africa and Asia. Leopards, jaguars, and tigers are solitary animals that do not live in groups. They usually stalk their prey slowly for a while and then cover the last few meters or feet with a sprint or with one mighty leap. Leopards and jaguars sometimes lie on a branch and drop on to prey that walks underneath. They kill by biting their prey in the neck

JAGUARS AND LEOPARDS

Cats are good climbers and spend a lot of time in the trees. Their coats provide excellent camouflage among the sun-dappled leaves. Jaguars are most common in dense forests. They catch monkeys, sloths, and birds in the trees, but most of their hunting is done on the ground, where they catch deer, peccaries, and capybaras. Peccaries are pig-like animals, while the capybara is like an enormous guinea-pig. Leopards eat many kinds of animals, including antelopes, monkeys, and pigs.

▶ Leopards are usually yellow with black spots like the jaguar, but in the wettest rainforests many of them are black and their spots are hardly visible. These leopards are often called panthers or black panthers.

▼ Well camouflaged by its spotted coat, the jaguar watches and waits for the right moment to attack.

TIGERS

Rainforest tigers are smaller and darker than other tigers. Their stripes conceal them well in the thick vegetation that grows along the river banks where they hunt. Tigers search for food when they are hungry, stalking their prey slowly until they are within about 65 feet (20 meters) and then charging rapidly. The prey is often knocked down with one blow from a huge paw. Tigers will eat any animal that they catch, but deer, antelopes, and pigs are their main prey. They also kill and eat water buffalo and sometimes elephant calves.

▶ Tigers are good swimmers and often plunge into rivers after their prey.

EAGLES IN THE TREE-TOPS

The world's biggest eagles live in the rainforests. They are the main predators in the canopy. Eagles' wings are very wide but short. They also have long rudder-like tails. This gives the birds great agility, allowing them to chase, at high speed, their prey between the trunks and branches of trees. They catch monkeys and many other animals with their huge **talons**, and tear them to pieces with their hooked beaks. Eagles have excellent eyesight. They spot their prey while soaring or while perched high in one of the emergent trees that stand high above the canopy.

CROWNED HEAD

Crowned eagles are named for the crest, or crown, of feathers at the back of their head. They live in Africa, in rainforests and in open country. Crowned eagles kill forest antelopes on the ground as well as monkeys in the tree-tops.

▲ The crowned eagle, with its crown folded down, is alert for any movement that might indicate a meal.

▼ This white-throated capuchin monkey, reaching for water in a tree hole, must always be on the look-out for harpy eagles ready to scoop it up in their talons.

The crowned eagle weighs about 8.6 pounds (4 kg) and can lift prey many times larger. Prey that is too heavy to be carried is torn apart and hidden in the trees to eat later.

BIGGEST EAGLE

The harpy eagle is the biggest of all eagles. Up to 43 inches (110 cm) long and weighing up to 18 pounds (8 kg), it feeds mainly on monkeys and sloths, and sometimes spiny porcupines. The harpy eagle's talons are as big as a large human's hands and it can plunge into the canopy to snatch its prey at a speed of 50 miles an hour (80 km). It lives in tropical South America. The New Guinea harpy eagle hunts in or below the forest canopy. This very rare bird usually sits on a branch and flies out to chase other birds. It also plunges to the ground to catch young pigs and other mammals. A good runner, it even chases its prey on the ground.

▼ Also known as the monkey-eating eagle, the Philippine eagle is one of the rarest birds of prey. It is almost as big as the harpy eagle. It catches monkeys and it sometimes drops down to snatch small deer from the forest floor.

❖ GLOSSARY ❖

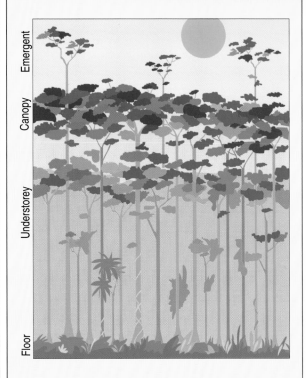

▲ **A rainforest has layers of vegetation: fungi and low-growing plants on the forest floor, slender, young, branchless trees that form an understorey below the vast, dense canopy of tree-tops. Taller trees called emergents, sometimes poke their heads through the canopy.**

Agile Quick and nimble.

Amazon A great river in South America and the area around it, which contains the world's largest rainforest. The Amazon is the second-largest river in the world at 4080 miles (6570 km). It runs through Peru and Brazil.

Amphibian Any member of the group of backboned animals that includes frogs, toads and newts. Most of them grow up in water and live on land when they mature. The name means 'double life'.

Buttress roots Large roots that spread from the base of a tree like low walls and help to support the trunk.

Camouflage Skin colors and patterns that help an animal blend with its surroundings and avoid the attention of predators.

Canopy The 'roof' of the rainforest, formed by the leafy branches of the trees. It is usually about 98 feet (30 meters) above the ground and it cuts off most of the light from the forest floor.

Carnivore An animal that feeds mainly on meat, or flesh.

Carrion The meat of animals that are already dead.

Cold-blooded A cold-blooded animal is one that cannot maintain its body at a constant temperature. Reptiles and amphibians are cold-blooded. Their body temperatures are usually similar to those of the surroundings, so they are not necessarily cold. In tropical rainforests, cold-blooded animals can actually be very hot.

Constrictor The name given to any snake that kills by wrapping its body around its prey and squeezing it until it cannot breathe.

Domestic Made tame for the household.

Emergent Any large tree that grows above the rainforest canopy.

Extinction The total disappearance of a particular plant or animal species from the earth, often brought about when people destroy forests and other habitats and leave the animals nowhere to live. Many rainforest species are in danger of extinction.

Food chain A sequence of plants and animals that feed on each other and pass energy along the chain. An example is flower – fly – spider – bird – cat. There are rarely more than five links in a chain and each one starts with a plant. Food chains in a particular habitat are all linked together into a complex food web.

Habitat The natural home of a plant or animal species. It may be a whole forest or just a tree trunk, or even a pool of water trapped by a plant.

Herbivore An animal that feeds on plants.

Invertebrate Any animal without a backbone, including insects, worms, and spiders.

Larva (plural larvae) Stage in the development of some animals. Caterpillars and tadpoles are larvae.

Leaf litter The layer of dead and decaying leaves on the forest floor.

Mammal Any member of the class of animals in which the females feed their babies with milk. Mammals are warm-blooded and most have hair or fur. Examples include human beings, monkeys,

ENDANGERED!

RAINFORESTS are home to more plants and animals that any other habitat on earth. They are important to the world but they are in danger of destruction. Many of the animals and plants shown in this book are endangered. Their rainforest habitat is slowly being destroyed by humans. If you are interested in knowing more about rainforests and in helping to conserve them, you may find these addresses and websites useful.

Friends of the Earth
USA - 1025 Vermont Ave NW, 3rd floor,
Washington, DC, 20005-6303
Canada - 47 Clarence St. Suite 306,
Ottawa, ON, K1N 9K1

Rainforest Foundation, U.S.
270 Lafayette Street, Suite 1107
New York, NY, 10012 USA

Rainforest Alliance
65 Bleecker Street, New
York, NY,
10012 USA

Rainforest Action Network
221 Pine Street, Suite 500
San Francisco, CA
94104 USA

Greenpeace
USA- 1436 U Street NW
Washington, DC, 20009, USA
Canada - 250 Dundas Street West, Suite 605
Toronto, ON, M5T 2Z5, Canada

Rainforest Alliance
http://www.rainforest-alliance.org

Friends of the Earth
http://www.foecanada.org
http://www.foe.org/FOE

Environmental Education Network
http://envirolink.org.enviroed/

Greenpeace
http://greenpeace.org

◀ **The map shows the location of the world's main rainforest areas.**

NORTH AMERICA
EUROPE
ASIA
Tropic of Cancer
AFRICA
Equator
Tropic of Capricorn
SOUTH AMERICA
AUSTRALIA

deer, cats, and elephants.

Mangrove swamp A coastal swamp covered with evergreen trees whose roots form dense tangles.

Nocturnal Active by night.

Omnivore An animal that eats both plant and animal matter.

Oxygen The gas in the air and in water that enables animals to breathe.

Predator Any animal that hunts and kills other animals for food.

Prey Any animal that is killed by another for food.

Queen The name given to the egg-laying female in a colony of social insects, such as ants or termites.

Reptile Any member of the group of backboned animals that includes tortoises, crocodiles, snakes, and lizards. They have scaly skins.

Saliva The juice in the mouth of an animal that lubricates food and begins the digestion process.

Scavenger An animal that feeds mainly on dead and decaying matter.

Suffocate To smother or strangle to death.

Talon A sharp claw, especially of a bird of prey.

Tropical Describes the tropics – the warm areas around the equator.

Understorey The layer of vegetation growing below the canopy, consisting mainly of young trees.

Venomous Having venom or poison that can be fired at or injected into another animal. The venom is used for defence or to kill prey.

Warm-blooded Warm-blooded animals keep their bodies at a constant high temperature, no matter what the surrounding temperature may be. Birds and mammals are warm-blooded animals.

❂ INDEX ❂

S0-AFA-697

To

NATALIE,

From

....

You may have heard the stories?
I tell you they are true!
A superhero lives nearby.
But where? I wish I knew!

She's got an "N" upon her suit,
Her cape is long and flowing.

Who is the girl behind the mask?
There's just no way of knowing!

They say she's very clever.
She's brave and fast and strong.
Do you know who she is yet?
Her name's SEVEN letters long!

SUPER DUPER CLUES

- She wears a mask
- She can fly—sort of!
- She's about this tall:
- Her name is N _ _ _ _ _ _

Oh no! A cat's stuck up a tree!
But who will get it out?
Our hero stops and says,
"I'll use my. . .

SUPER NATALIE SHOUT!!!"

This little girl is crying.
Her trike has got a flat.

I guess it must be hungry work
When one is fighting crime.
She runs and jumps and dives around.
She's moving all the time!

Super Natalie's so strong,
She simply can't be beaten.
It must be all the salad, sprouts,
And broccoli she's eaten!

There's trouble on the playground!
Kids start to scream and shout.

Super Natalie will know
Just how to work this out!

She's the world's best superhero.
And she's got a super cuddle!

The next time you're in trouble,
Or ever in harm's way, shout. . .

This superhero stuff's hard work,
And now she's very sleepy,
But Super Natalie's afraid—
Her bedroom looks so creepy!

Yet heroes do not run or hide
When they are feeling scared!
Instead they face their fears head on.
That's why she's come prepared!

She's caring and she's helpful,
Always doing awesome deeds.
Super Natalie's the hero
Everybody needs!